An Herbalist's Guide to
Growing & Using Violets

Kathleen Brown

CONTENTS

Introduction

Few flowers inspire as much rhapsodic praise as violets. Beloved for both their beauty and their sweet and gentle fragrance in the garden, they've also been used for centuries in cooking and as an ingredient in an eclectic range of products — from soaps and perfumes to medicinal home remedies. It's no wonder that AnneBelle Rice of Long Beach, California, treasurer of the International Violet Society, describes herself as being "passionate" about violets.

In today's commercial products, the natural fragrance of violet has been largely replaced with synthetic scents. Among home gardeners and crafters, however, violets have never lost their popularity. The flowers are often used as garden ornamentals. As the herbalist Maurice Messegue poetically put it, violets "show up like amethysts in the hedgerows." In addition, for centuries people have used the fresh flowers to make delicately fragrant potpourris. And because the dried flowers retain their beautiful colors (if not their fragrance), they've traditionally been used for making decorative bookmarks.

The violet is also popular among creative cooks, who use the edible flowers — often in a crystallized form — on cakes, puddings, ice cream, and candy. The flowers, fresh or candied, are also used as a colorful and peppery-tasting garnish for salads and main meals.

Which Violets Have Medicinal Properties?

Viola odorata (sweet violet) and *V. tricolor* (Johnny-jump-up) are the two species most commonly used in herbal medicine. For the most part, it doesn't matter whether you use *V. odorata* or *V. tricolor* for healing — they're nearly interchangeable.

There are some exceptions, however:

- *V. odorata* is considered a better expectorant than *V. tricolor.*
- *V. odorata* is recommended for chronic bronchial conditions.
- *V. odorata* is said to have more pronounced effects against tumors in the lungs, throat, breast, stomach, and intestines.

Growing Violets

The violet genus *(Viola)* is one of the more than 20 genera of the family Violaceae. There are more than 500 species of violets, and many more varieties. Most species of violets are native to the Northern Hemisphere, and nearly 20 percent are indigenous to the United States.

Violets are easy to recognize by their little flowery "faces" and leaves, which unroll from the base as they mature. The lobes stand more or less erect while remaining inwardly curled. The lowest of the petals has a spur, which contains the sweet perfume and nectar.

One of the most popular violets, renowned both for its aromatic beauty and its potent healing properties, is the sweet violet *(Viola odorata)*. This rhizomatous perennial has small, heart-shaped, green leaves. It grows to a height of 4 to 6 inches

Growing at a Glance

Plant Cycle:
Annual or perennial, depending on species

Soil Requirements:
Will tolerate most conditions but thrive in deep, rich, moist soil

Light Requirements:
Shade or semishade

Water Requirements:
Most species prefer moist soil

Spacing Requirements:
6 to 12 inches (15–30 cm)

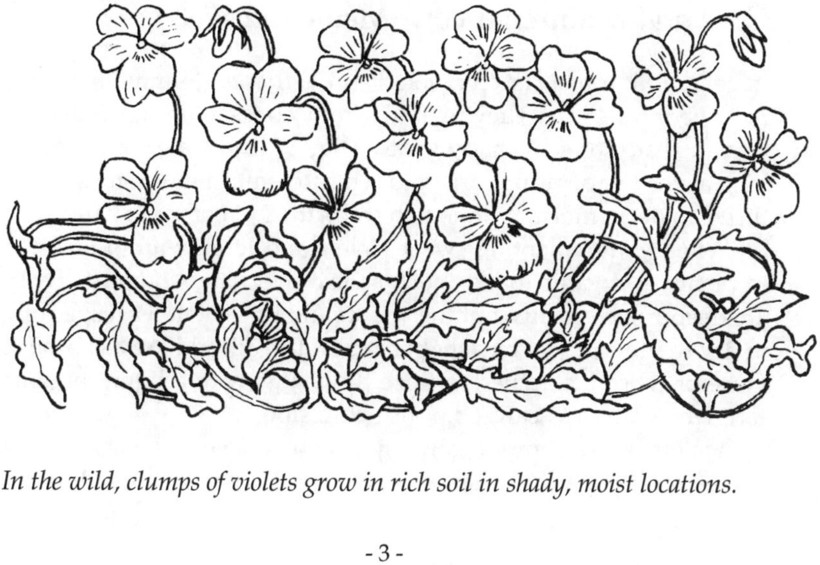

In the wild, clumps of violets grow in rich soil in shady, moist locations.

(10–15 cm) and has a spread of 6 to 12 inches (15–30 cm). The sweet violet blossom has three petals on its upper side and two below. The ½-inch (13 mm) or larger flowers range in color from white and pink to pale lavender, rose, and blue-purple. Sweet violets produce a rosette of foliage from which downy runners emerge and grow along the ground. They create a springtime carpet of lovely, sweet-smelling flowers, whether in the garden or growing wild. A clump of sweet violets can live between 8 and 14 years.

Violets make a wonderful living mulch under shrubs. In the south of France, where violets are grown for their fragrance, they are planted under orange, lemon, and olive trees. They are also excellent for naturalizing under many other trees, where nothing else wants to grow.

Many species of violets will grow even on fairly steep, shaded northern slopes, and they make a good weedproof ground cover. They also make good border plantings, especially when featured at the front of the garden. Violets are suitable for rock gardens, and some, such as pansies and Johnny-jump-ups, are excellent candidates for flower boxes or container planting. They look wonderful when grown with woodland plants such as primrose, cowslip, hellebore, wild strawberry, and lungwort. The sweetest smelling of violets, of course, must be included in any fragrance garden, along with other aromatic herbs, such as chamomile, pineapple sage, thyme, rose geranium, lavender, rosemary, marjoram, rose, and lemon balm.

Care and Maintenance of Medicinal Violets

In general, *V. odorata* (sweet violet) and *V. tricolor* (Johnny-jump-up) need very little care. They're hardy from Zones 4 through 8 and do best in shady or semishady spots. They will tolerate some sun as long as the temperatures are mild. They tolerate almost any soil but thrive in deep, moist, humus-rich soil with a pH of 5.5 to 7.0.

When you're planting sweet violets and Johnny-jump-ups, dig in plenty of organic matter, such as leaf mold and compost. These are deep-rooted plants, but the soil should be kept moist, especially in dry weather. If you have light or gravelly soil, add extra compost. Once or twice a summer, fertilize the plants with a liquid organic fertilizer, such as a seaweed or fish emulsion.

Violets spread prolifically by runners (see Propagation by Runners on page 7). Be sure to give them room to spread, or be prepared to divide them frequently. Under ideal growing conditions, violets may become invasive — although it's hard to imagine having too much of their fragrance and flowers!

When transplanting violets from the wild, try to duplicate their natural growing conditions as closely as possible. For example, if the plants were growing in a shady, moist area, try to locate them in a similar spot in your garden. It's generally a good idea to plant violets with other plants that enjoy similar growing conditions. These include angelica, comfrey, foxglove, horsetail, lady's-mantle, pennyroyal, sorrel, sweet flag, sweet woodruff, and valerian.

Violets through the Seasons

Here's an annual calendar of the basic maintenance that your medicinal violets will require:

- **In spring.** Sow seeds and put rooted runners into pots or replant them in the garden. You should also divide established plants. Cut back plants after their first bloom to encourage a second flowering.
- **In summer.** Deadhead the flowers to prolong the season. Renew mulch to keep plants cool.
- **In autumn.** Sow seeds for spring flowers.
- **In winter.** Put on a light, even mulch of well-rotted manure or compost.

By giving your violets just a little attention every season, you'll enjoy them for many years to come.

Propagation by Seeds

Violet seeds germinate more uniformly and vigorously if they've first been stratified, or subjected to a cold treatment. You can sow them in sandy soil in a protected area, such as an open cold frame, in fall or artificially stratify them in the refrigerator and plant in spring. To stratify in the fall, plant the seeds in a tray containing soil-based compost. Water, then cover the tray with a sheet of glass or plastic. Put the tray in a corner of the garden and cover with mulch or set it inside a cold frame. In the spring, transplant them to the spot you've picked out.

To stratify in the spring, 6 to 8 weeks prior to planting, set the seeds in a multiflat or pot filled with a damp mixture of potting soil and sphagnum peat moss, cover with a plastic bag, and let sit in the refrigerator. You can also simply refrigerate the seeds in a resealable plastic bag filled with damp sand or slightly damp sphagnum peat moss.

To stratify seeds in the spring, sow them in a multiflat, cover with plastic, and set in the refrigerator for several weeks.

In warmer regions, sow seeds outdoors in early spring, covering them with just enough soil to eliminate light. In colder areas, start the seeds indoors in trays or multiflats 10 to 12 weeks before transplanting outdoors. Place them in a dark closet or cover the trays with an opaque material, such as aluminum foil, until germination takes place. Then remove the cover. Harden off the seedlings and then transplant outside, up to a month before the last frost or in late autumn, for fresh color until frost. Seeds take 1 to 3 weeks to germinate. Space or thin the plants so that they are growing 6 to 12 inches (15–30 cm) apart.

When grown from seeds, violets may not flower until their second year.

After they've been hardened off, seedlings started indoors can be transplanted to the garden.

Saving Seeds

Sweet violets and other "stemless" violets produce two types of flowers. The showy, fragrant blossoms of spring are almost always sterile and rarely produce seeds. In summer, however, small, pod-shaped green flowers form at the base of the plant. These petal-less blossoms never open and are self-fertile. In late summer, the tiny pods change from green to reddish brown and pop open, spreading seeds over the ground.

To save seeds, collect the pods after they have changed color but before they open. Put the pods in an open, dry glass jar until they split, discharging their seeds. Then place the seeds on a sheet of clean white paper for sorting. Store them in a paper envelope or a dry glass jar in a cool, dark location, where they'll remain viable for up to a year.

In many species of violets, the seeds are carried in small, pod-shaped flowers. When these pods turn from green to reddish brown, they are getting ready to release their seeds.

Propagation by Runners

Many species of violets, including sweet violets and Johnny-jump-ups, produce runners — long creeping stems that root at various places along their length. Propagating violets from runners is very easy and can be done in either autumn or spring. In northern climates where winter comes early, however, it's best to propagate violets by runners in spring.

When you're taking runners in fall, gently remove them from the parent plant. Plant them in cell trays with a bark-and-peat soil. Their root systems will develop over the next few weeks.

When propagating violets from runners in spring, plant them directly into the ground, making sure that the base of each crown (the point from where the leaves rise and the roots form) is flush with the soil. It's a good idea to plant three crowns together for a better show and as insurance against loss.

Propagation by Divisions

Violets bloom better when they're divided every year. The best time to divide is early spring, when the violets are just beginning their fresh burst of growth. A division that's made in spring will usually become full sized in one season.

Step 1. Start by loosening the soil around the perimeter of the plant's root system. Then spade under the plant and lift it up. Set the clump on the ground.

Step 2. Divide the plant into smaller clumps. Each division should have its own root, stem, and leaves.

Step 3. Plant each clump in a hole or pot filled with well-rotted compost at approximately the same depth at which the plant was growing before. Water with a diluted solution of organic fertilizer to minimize transplant shock, and continue to water well until the plants are established.

You can also divide violets in summer, but this requires more careful handling. Provide a slow, deep watering before dividing the plants. Divide them early in the day or in the evening so the sun won't stress them. Keep as much soil around the roots as possible, and plant the divisions immediately. If you can't plant right away, wrap them in damp newspaper and place in plastic bags for moisture retention.

Container Gardening with Violets

Violets grow well in containers. Plant them in a mixture of bark, peat, and sand, and apply a fertilizer after they flower. If you're bringing your violets inside for winter, be aware that they don't like heat, especially not dry heat. Keep them in a cool place, where temperatures don't rise above 45°F (7°C). They need good air circulation as well as regular watering.

Pests and Problems

According to folklore, loud thunder weakens violets and causes them to be spotted with stars. This is unlikely, of course, but violets can be temperamental, and gardeners who raise them face several potential problems. Slugs and snails are common because of the moist growing conditions in which violets thrive. These pests chew holes in the leaves and flowers, sometimes devouring them completely. You can control slugs and snails by sprinkling diatomaceous earth around the base of the plants. Many gardeners make slug and snail traps by putting out shallow dishes filled with beer. The critters are attracted to the beer and drown when they fall into the dishes.

The major pest, especially for container plants, is the red spider mite, which is common on overly stressed plants. These spider relatives are less common in soils that are sandy or have a pH below 5.7. Infested plants will be stunted and less vigorous, and their leaves are often peppered with minute yellowish spots. You can combat mites by spraying the leaves with insecticidal soap.

Yet another potential problem is black root rot, caused by a fungus that is most common in plants grown in soil with insufficient drainage. Symptoms of black root rot include blackening of the roots due to decay. In young plants and seedlings, the stems can collapse, resulting in a condition called damping off. Once your plants contract black root rot, there's not much you can do to save them.

Finally, even though violets don't like hot sun, too much shade can prevent them from flowering.

A Gardener's Secrets

AnneBelle Rice of the International Violet Society is a practiced grower of violets. Some time ago she decided to find out which violets would be hardy in southern California. She obtained 50 species and varieties from around the world and has trimmed the list to 25 that have proved themselves worthy.

One problem Rice has encountered is the European midge gall, a pest that stunts the plant's growth and causes its leaves to thicken. "There is no control except picking off the leaves," she says. (Don't toss the affected leaves in the compost pile!) She has begun experimenting with row covers to see if they will provide protection.

Rice has some good advice for gardeners in warm climates: "During hot spells, spray the foliage to cool the plants. They can be kept fairly dry after they are established, as long as they are in the shade and planted with a moisture-retentive mulch. Water well to encourage the roots to go deep into the soil."

For optimum bloom in warm climates, she adds, violets need sun in winter, and growers should be sure to cut back the foliage in fall to encourage blooming. "Violets will naturalize without any extra effort on your part," she says. "So plant them everywhere you desire color, fragrance, and beauty."

Species and Varieties of Violets

An entire book could be written about the different species and varieties of *Viola.* Since I can't do them all justice here, I'll touch on just a few of the most popular.

- *Viola odorata* and *V. tricolor* are the species believed to have the most potent healing properties, and their effects are very similar (see box on page 2).
- *V. odorata* is grown for its beauty, fragrance, and healing properties. It likes moist soil and shaded, woodland locations and is hardy in Zones 4 through 8.
- Two cultivars of *V. odorata* that produce stronger fragrances are 'Royal Robe', with deep purple blossoms 1½ inches (4 cm) across, and 'Rosea', whose blossoms have a dusky rose color and an intense aroma. Both are hardy in Zones 4 through 8.
- *V. odorata* includes a wide range of color combinations. 'Coeur d'Alsace' has pink flowers; 'Princess of Wales' has large, round, blue flowers; 'Charm' has white flowers; and 'Royal Elk' has violet flowers.

Sweet violet
(Viola odorata)

- *V. tricolor,* also called Johnny-jump-up, is a biennial grown primarily for color. It grows 3 to 8 inches (8–20 cm) high. Unlike most violets, it enjoys sunshine and open meadows, and although it prefers well-drained loam, it will tolerate just about any kind of soil. It is hardy in Zones 3 through 8.
- *V. alba* has white flowers tinged with purple. It's a prolific producer of seeds and is hardy in Zones 4 through 8.
- *V. arvensis,* also known as the field pansy, grows 2 to 4 inches (5–10 cm) high. It has predominantly white to yellow flowers and is hardy in Zones 4 through 8.
- *V. canina* is the dog violet. It has blue-violet flowers and is hardy in Zones 6 through 9.
- Some violets have no scent at all. Two species, *V. labradorica* (hardy in Zones 2 through 8) and *V. selkirkii* (hardy in Zones 3 through 5), are often cultivated in hedges and woodland settings where a natural appearance, rather than fragrance, is the primary goal.
- *V. lutea* is the mountain pansy. It grows 3 to 8 inches (8–20 cm) high and has single-colored flowers, usually yellow, blue, or violet. It is hardy in Zones 4 through 8.

Common pansy
(Viola x wittrockiana)

- *Viola* x *wittrockiana,* better known as the common garden pansy, is a favorite of many gardeners. There are innumerable varieties — in many sizes and colors — to choose from. Pansies are usually grown as annuals or biennials.
- *V. pedata,* or bird's-foot violet, is one of the most common kinds of violet and grows up to only 2 inches (5 cm) in height. Its flowers are white or sometimes bicolored, featuring dark violet upper petals and pale lilac lower petals. It's hardy in Zones 4 through 9.

Bird's-foot violet
(Viola pedata)

Harvesting Time

As always, it's better to harvest plants from your own garden than from the wild. However, if you decide to "wildcraft" violet leaves, it's important to practice ethical wildcrafting. Harvest only a few leaves from each plant, and be sure to clip plants above the ground so as not to disturb the roots.

When harvesting violets for cooking, gather blossoms and young, tender, bright green leaves. Wash them thoroughly to remove dirt, then snip the leaves into ¼-inch (6 mm) shreds with kitchen shears.

Medicinally, the entire plant is used to treat a number of ailments, including headache, insomnia, and constipation. (Violet contains salicylic acid, which is a relative of modern-day aspirin.) The roots are a much stronger expectorant than the other parts of the plant; however, the roots also contain higher amounts of the alkaloid violine, large amounts of whichcan cause vomiting and diarrhea.

Flowers

When they are fully open, pick sweet violet flowers early in the day. Keep the stems in water until you're ready to use them. A tiny container of cut flowers will fill a room with fragrance. Don't think that you're harming the plants by harvesting their flowers: The more you pick, the more the plant produces.

Dry the flowers in the same manner in which you dry leaves (see the description on the facing page). However, if you're going to be drying the flowers, keep them away from light as much as possible because exposure to light will cause them to give up their scents along with their colors.

Roots

The roots contain most of the active ingredients that make violets such a popular herbal remedy. These compounds will be most active when you dig violets in autumn.

Step 1. Dig up the plants. Using a sharp knife, snip the roots from the stems. (Save the leaves and flowers for drying because they contain many of the same healing properties as the roots.)

Step 2. Once you've harvested the roots, cut any pieces larger than 1 inch (2.5 cm) in diameter into smaller pieces to avoid mold growth during the drying process. Thoroughly wash the roots, then pat dry.

Step 3. Hang or lay the roots on screens in a well-ventilated area away from direct sunlight. Depending on their size, the roots should dry in a week to several weeks.

Step 4. When completely dry, store the roots in tightly covered glass jars in a cool, dark place. Be sure to label and date your harvest so you can monitor its freshness. The roots should keep for about a year.

Leaves

Leaves from violet plants contain many of the same healing properties as the roots, although in lesser amounts. They should be harvested in summer.

Step 1. With a sharp knife, clip the leaves from the stems of the plants.

Step 2. Lay the leaves on screens and let them dry for about seven days, turning them regularly so they don't mat and mold.

Step 3. Once they are dry, place the leaves in an airtight glass container and store in a cool, dry location, where they'll keep for up to a year.

Quick Tip

To follow the French tradition, sprinkle freshly picked violet blooms and leaves over salads of young greens. The leaves are slightly peppery, so you don't want to use a lot of them.

Storage Tips

Dried violet leaves, roots, and flowers will store well, but you must take a few precautions in order to preserve their potency:

- Roots must be completely dry before they're stored; otherwise, mold will form.
- Store dried roots, leaves, and flowers in airtight glass containers. Dark glass is especially good because it prevents light from fading the herbs and reducing their medicinal potency. If you do use clear glass, be sure to put the jars in a dark place, or at least out of direct light.
- Store roots, leaves, and flowers immediately after drying. This will help preserve their volatile oils, color, texture, and flavor.
- Label the containers carefully, noting the date they were stored. This will allow you to monitor freshness.
- Check the containers regularly to make sure there is no condensation. If there is any moisture, remove the roots and check them for signs of mold or any deterioration. If they seem all right, allow them to dry out further and then restore them to a dry, clean, airtight container.

Are Violets Really Medicine?

Violets have a long history of healing. In ancient times, Hippocrates classified the violet as a "moist" plant, best used for treating liver disorders as well as bad tempers. As seventeenth-century herbalist Culpepper put it, "All the violets are cold and moist while they are fresh and green, and are used to cool any heat or distemperature of the body." Culpepper himself recommended using violet remedies for treating a variety of eye and skin disorders, as well as for constipation and congestion. As far back as the 16th century, the English used syrups made from violets as mild laxatives for children. Adults took violet remedies for epilepsy, pleurisy, and jaundice.

Many of the traditional medicinal powers of violets have undoubtedly been exaggerated, and a long history of folklore use isn't the same as scientific evidence. However, scientists today have found that violets — the flowers as well as the roots and leaves — do contain a variety of healing compounds.

Research has shown, for example, that *Viola odorata* contains, among other things, phenolic glycosides, saponins, flavonoids, alkaloids, tannins, and mucilage. These compounds act in a variety of ways to promote healing. For example:

- The flavonoids act as diuretics, which could be helpful for people with elevated blood pressure.
- The alkaloids may affect blood pressure because they have vasodilating effects — that is, they cause the blood vessels to relax, allowing blood to flow more easily.
- The saponins and mucilage act as soothing expectorants, which can help relieve harsh, irritating coughs and other bronchial problems.
- Violets contain salicylic acid, a painkilling and anti-inflammatory compound that's similar to the active ingredient in aspirin. This might be helpful for people who are trying to reduce the discomfort of arthritis or other joint problems.
- According to herbalist and plant expert James A. Duke, Ph.D., violet flowers contain significant amounts of rutin, a compound that seems to significantly strengthen the capillaries. For people with inflammation, this means less "leakage" from the blood vessels and less swelling and pain.
- Finally, violets are rich in the antioxidant compounds beta carotene and ascorbic acid (vitamin C). A ½-cup (140 ml) serving of violet leaves provides as much vitamin C as four average-sized oranges and more than the USDA Recommended Daily Value for vitamin A (beta carotene is converted to vitamin A in the body).

In Germany, an expert panel called Commission E, which judges the safety and effectiveness of herbal medicines, has approved violet tea as a remedy for some skin problems. And herbalists around the world have found that violets are, in fact, helpful for treating a wide variety of health problems.

The Cancer Question

As far back as 500 B.C., poultices made from fresh violet leaves were used to treat skin cancer. Much later, in 1901, the *British Medical Journal* discussed the violet's cancer-fighting potential. And many herbalists today have found violets useful for treating a number of cancers. Although it's always tempting to dismiss claims about the ability of herbs to block cancer and other serious illnesses, in the case of violets, there's some science to back up the claim. In 1960, for example, animal studies showed that extracts of violet were able to damage tumors. Further, there were reports that violets were successfully used to relieve the pain of cancer. Catherine Booth, wife of the founder of the Salvation Army, reputedly used violet leaves for this purpose.

Because violets contain mucilage, which seems to relieve inflammation and irritation in the stomach and intestines, they have been used in Britain to treat stomach as well as breast cancer. It's important to note, however, that the British Pharmacopoeia does not uphold this treatment. There simply hasn't been enough study in this area to say for certain what effect, if any, violets have on the growth and development of cancerous tissue.

A Fragrant Medicine Chest

Despite the dearth of scientific evidence for their medicinal effects, violets are very widely used as healing herbs. It's interesting to see just how many conditions they've been used for — not only in the past, but today. According to herbalists, violets may:

- Detoxify the body
- Eliminate lymphatic congestion
- Relieve eczema and other chronic skin problems
- Regulate immunity
- Reduce inflammation
- Ease constipation
- Relieve itching
- Work as a mouthwash
- Strengthen blood vessels
- Improve the urinary system
- Relieve incontinence
- Ease the discomfort of menopause
- Relieve colds and flu
- Reduce fatigue and stress
- Lower blood pressure
- Promote healing of wounds

Cautions and Concerns

It's important to remember that violets are commonly used both as a laxative and to induce vomiting. Therefore, whatever the condition being treated, you need to be aware violets may have some uncomfortable side effects.

Professional singers may have a unique issue to contend with: The essential oils in fresh violets and in violet-based perfumes, in some cases, cause the vocal cords to swell. You may not want to give a violet bouquet to your favorite singer before a performance!

The main danger isn't so much from the violets themselves as it is from accidentally picking the wrong plant when harvesting. Unless you're an experienced wildcrafter, it's best to harvest violets only when the plants are in flower. Otherwise, the heart-shaped leaves may be confused with those of other plants — which may be toxic.

Caution

Keep in mind that African violets are not related to *Viola* spp., and they are not edible. Do not substitute African violets for violets in any recipe!

Making Your Own Herbal Remedies

Now that you are familiar with violets and their medicinal uses, you're probably ready to start making some of your own formulas. In the following pages, you'll find all the information you need to make basic infusions — the most common type of remedy — as well as tinctures and infused oils.

Quick Tip

Violets can discolor cooking utensils. To prevent this, use nonreactive pots made of stainless steel, ceramic, or glass when preparing recipes.

Making an Infusion or Decoction

There are two ways to prepare herbal teas: You can make an infusion, in which the herb is steeped, or you can make a decoction, in which the plant matter is simmered over time.

Infusions. Infusions are best when you're using leaves, flowers, or seeds. These ingredients easily release their essential oils when they're steeped in hot water — and they easily lose their value when they're simmered. Pour 1 cup (235 ml) of boiling water over 1 to 2 teaspoons (5–10 ml) of dried herbs, or 2 to 4 teaspoons (10–20 ml) of fresh. Cover and let steep for at least 15 to 20 minutes, then strain well.

Decoctions. Decoctions are made by simmering roots, bark, and other woody parts of the plant, then cooling the liquid. The simmering is necessary to extract the valuable properties from these fibrous materials. To decoct a cup of tea, combine 2 teaspoons (10 ml) of dried herb and 1 cup (235 ml) of water in a saucepan. Cover, bring to a boil, then reduce the heat and simmer for at least 20 minutes and up to an hour. Strain the herbs and enjoy.

Combinations. When you're making a tea with roots and leaves, you'll both infuse and decoct: Simmer the roots, covered, for at least 20 minutes; remove the pot from the heat and stir in the leaves; then cover and steep for at least 15 minutes.

Making a Tincture

A tincture is made by using a solvent — usually alcohol — to extract a plant's beneficial components. Most people make their tinctures with vodka, but if you're using particularly bitter herbs, you may want to disguise their taste by using brandy or rum. Nonalcoholic tinctures can be made with apple cider vinegar or glycerin as the solvent.

BASIC TINCTURE

Step 1: Process the herbs. If you're using fresh herbs, coarsely chop or mince them. If using dried herbs, powder them in a coffee grinder or with a mortar and pestle. In both cases, what you're trying to do is expose the plant's cell walls to the alcohol.

Step 1

Step 2: Immerse the herbs in liquid. Put the processed herbs in a widemouthed jar. The herbs should make up about half of the total volume. Then cover with twice as much liquid — apple cider vinegar, glycerin, or alcohol, such as vodka or brandy. (Many herbalists prefer brandy when making tinctures for throat, lung, and digestive conditions.) If you're tincturing with alcohol, use at least 80 to 100 proof (40 to 50 percent alcohol).

Quick Tip

If you're tincturing with dried herbs and glycerin, dilute 2 parts glycerin to 1 part water. Use the glycerin full strength with fresh herbs.

Step 3: Seal the jar. After adding the liquid, stir well, then seal the jar tightly. If you're tincturing with vinegar, be sure to cover the top of the jar with plastic wrap before putting on the lid. Otherwise, fumes from the vinegar will corrode the lid, making it difficult to open. Then set the jar in a dark place and let sit for 3 to 6 weeks, shaking occasionally.

Step 4: Strain and bottle the liquid. Strain out the herbs, using a layer of cheesecloth or a strainer lined with an unbleached coffee filter, and decant the liquid into smaller bottles (glass bottles with eyedroppers are ideal, especially if made of dark glass). Store in a cool, dark place, where the tincture will remain viable for up to 5 years.

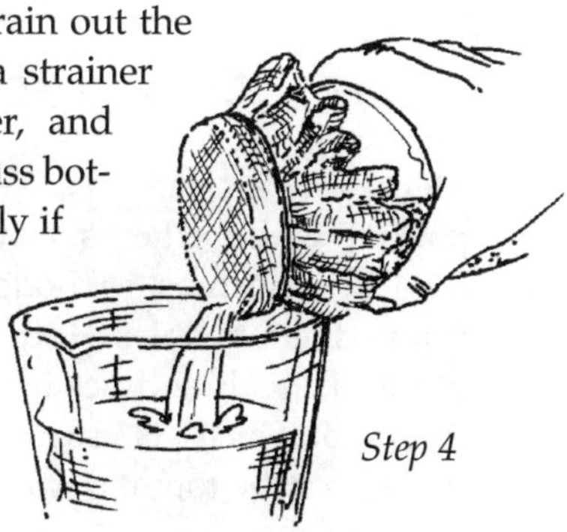

Step 4

Making an Infused Oil

Making an infused oil requires placing flowers and leaves in oil and allowing them to steep for a while. When using flowers to make the oil, be sure that they're fresh (but wilted, so that they don't retain too much moisture — see the box on the next page), because once they're dried and stored, flowers lose their effectiveness. If you do use dried herbs, make sure no moisture is present. Otherwise, molds or even botulism could grow.

Infused oils are often used as an astringent or as a mild antibacterial that can be applied to any minor wound or skin irritation. Some people use the oils for massage or as an ingredient to be added to ointments. Violet-infused oil, in particular, is both edible and aromatic; it can be used to make oil-and-vinegar salad dressings and makes a wonderfully scented base for healing salves.

Start with top-quality extra-virgin olive oil if you intend to use the oil for medicinal purposes. Oils such as apricot, almond, and grape seed are better for cosmetic purposes. There are two techniques to choose from: The sun method and the stovetop method.

Wilting Herbs

Infused oils are best made with dried herbs because excess moisture can create conditions in which mold or even botulism grows. However, some herbs, especially flowers, lose medicinal potency when dried. In such cases, you can make an infused oil from the freshly wilted herb. Once wilted, most of the plant's moisture content will have evaporated. Simply lay the fresh herbs out on a screen or paper towels in a place with plenty of air circulation and away from direct sunlight. After 24 hours, they should be nicely wilted and most of the moisture they contain should have evaporated.

THE SUN METHOD

Step 1. Fill a clean, dry, widemouthed glass jar to the top, loosely packed, with the herbs. Cover with the oil and stir with a nonmetal utensil — such as a wooden spoon or chopstick — to release any trapped air bubbles. Top off with more oil, seal, and set in a warm, dry spot, such as a sunny windowsill or the top of a water heater, for 2 to 6 weeks.

Step 2. When the oil has turned the color of the herb(s) being infused, press the oil through cheesecloth to filter out the spent plant matter. Then let the oil stand so any water will separate. Pour off the water and store the oil in a sealed container in a dark, cool place.

THE STOVETOP METHOD

Step 1. Place the herbs in the top part of a double boiler, then cover with oil. For a yield of 2 cups (470 ml), use 2 cups (470 ml) of dried herb or 4 cups (950 ml) of freshly wilted herbs to 4 cups (950 ml) of oil. Completely immerse the herbs and simmer slowly over water for about 3 hours. Don't let the oil bubble or smoke. Long, slow cooking produces the best results.

Step 2. Strain the oil by lining a wire strainer with muslin or a coffee filter. Place the herbs in the strainer, then press to remove as much liquid as possible.

Step 3. Bottle the liquid. If the oil is to be ingested, store it in the refrigerator. If it's to be used only topically, it does not need to be kept in the refrigerator; just store it in a cool, dark place.

Making a Poultice or Plaster

A poultice is a warm, moist mass of powdered or crushed fresh herbs. It's applied directly to the skin to relieve inflammation and other skin conditions. A plaster is much like a poultice, the only difference being that the herbs are placed between two thin pieces of cloth instead of applied directly to the skin.

To make a poultice, chop or shred the herbs, then moisten them with hot water, witch hazel, tea, or tincture. Apply directly to the skin.

Violet Recipes

Now it's time to take a look at some of the ways to prepare this wonderfully versatile herb. In the following pages, you'll find recipes for violet teas and tinctures, violet garnishes for salads, violet vinegar, and even violet mousse. Plus, there's even a recipe for crystallizing violet — a natural way to get extra vitamin C!

VIOLET TEA

Most teas have a dull, brackish appearance, but not this one. Unlike most flowers, violets readily give up their colors when they're steeped in hot water, making violet tea very beautiful to look at. Herbalists recommend violet tea for sore throats because it acts as a mild sedative and has a slight anesthetizing effect and because mucilage from the flowers coats the throat and relieves irritation. Violet tea can also be used as a gargle, a mouthwash for sore gums, and a wash for rashes.

1 handful violet leaves
1 cup water

Follow the instructions for making an infusion on page 17. Drink ½ cup (120 ml) every 2 to 3 hours.

TONIC TEA

When you feel a cold coming on, this tea will help you feel better generally and will help relieve some symptoms, such as congestion.

- **1 part borage leaves and flowers**
- **1 part dandelion flowers**
- **1 part elderberries**
- **1 part nettle leaves**
- **1 part parsley leaves**
- **1 part rose hips**
- **1 part rosemary leaves**
- **1 part sage leaves**
- **1 part salad burnet**
- **1 part violet leaves and flowers**
- **1 part sweet woodruff leaves**
- **1 part yarrow leaves**

Follow the instructions for making an infusion on page 17. Drink 1 to 3 cups (235–700 ml) a day.

MENOPAUSE FORMULA

This herbal blend will help support and balance the body's hormones. It's good for hot flashes and other types of menopausal discomfort and also seems to boost energy.

- **1 part dandelion root**
- **1 part dong quai root**
- **1 part fenugreek seeds**
- **1 part wild yam root**
- **1 part alfalfa leaves**
- **1 part chickweed leaves**
- **1 part horsetail**
- **1 part motherwort leaves**
- **1 part nettle leaves**
- **1 part red clover blossoms**
- **1 part red raspberry leaves**
- **1 part violet leaves**

Decoct the dandelion, dong quai, fenugreek, and wild yam, following the instructions given on pages 17–18. Then add the remaining herbs and steep, covered, for 15 to 20 minutes. You can also use this remedy in tincture form; see the instructions on pages 18–19.

SPRING FLING TEA

This is a light, refreshing tea. Many people use it as a palate-cleansing follow-up to the heartier "spring cleansing" teas. To maximize the fresh, lively taste, it's best to use fresh herbs rather than dried.

- **2 parts violet leaves and flowers**
- **2 parts sweet cicely leaves**
- **1 part nettle leaves**
- **1 part raspberry leaves**
- **1 part mint leaves**

Mix 10 to 12 tablespoons (150–180 ml) of fresh herb, or 5 to 6 tablespoons (75–90 ml) of dried, in 2 to 3 cups (470–700 ml) of cold water and let sit for several hours. Drink warm or cool.

Variation: When you're in a hurry, mix the herb in the water, cover, and slowly heat until almost boiling. Then remove from heat and let steep, covered, for 5 to 10 minutes before drinking.

DETOX AND SUPPORT

This blend is best used as a tincture. It supports the liver and various glands, helping the body eliminate toxins more efficiently.

- **2 parts red clover blossoms**
- **2 parts calendula flowers**
- **2 parts cleavers leaves**
- **1 part licorice root**
- **1 part violet leaves**
- **1 part dandelion root**

Make a tincture with the ingredients, following the instructions on pages 18–19. Take 25 to 50 drops of tincture two to three times a day. The treatment is most effective when continued for 3 months.

Quick Tip

Tinctures contain such concentrated essences of the plant that they often taste very bitter. To make them more palatable, mix the dose with a small amount of hot water, tea, or juice, then drink.

DECREASE CYSTS

This formula, best used as a tincture, improves lymphatic drainage, which can help prevent cysts from forming in breast tissue. It also helps support and balance the body's glands.

2 parts calendula flowers
2 parts violet leaves
2 parts vitex berries
1 part cleavers leaves
1 part dong quai root
1 part astragalus root

Follow the instructions for making a tincture given on pages 18–19. Take 25 to 50 drops of tincture three times a day for 2 to 3 weeks.

MINERAL-RICH TEA

In today's hectic world, it can be difficult to find time to eat as well as we should. This formula is packed with healthful vitamins and minerals and makes a great daily supplement.

2 parts borage leaves and flowers
2 parts oatstraw
2 parts red clover blossoms
1 part violet leaves and flowers
1 part chamomile flowers
1 part hawthorn flowers and berries
1 part horsetail
1 part nettle leaves
1 part raspberry leaves

Make an infusion of the ingredients, following the instructions on page 17. Drink 1 to 3 cups (235–700 ml) a day, hot or cool.

Violet Eyewash

If you'd like to use violet for soothing sore eyes, make a violet infusion following the instructions on page 17. Let the liquid cool to room temperature, then use it to flush and bathe the eyes. Refrigerate any leftovers and use within a week.

Some herbalists, incidentally, believe that eating violet petals and leaves will improve vision.

BLADDER RELIEF

Nearly every woman will eventually have to deal with a painful bladder infection or cystitis attack. This recipe will provide quick relief.

1 part juniper berries, crushed
1 part bearberry (uva ursi) leaves
1 part violet leaves
1 part goldenrod leaves

Make an infusion with the ingredients, following the instructions on page 17. Drink 1 cup (235 ml) six times a day during the acute phase of cystitis.

MIGRAINE TEA

Many healers believe that a plant's appearance and habits reflect its healing abilities. Sweet violet hides from the sun and loves shade — just as many people do when a migraine is coming on! Plus, this tea has the advantage of working very quickly.

Rosemary is an essential ingredient in this recipe because it dilates the blood vessels, promoting blood flow.

6 parts rosemary leaves
4 parts violet leaves
4 parts peppermint leaves
4 parts lemon balm leaves
3 parts feverfew leaves
½ part violet flowers

Make an infusion with the ingredients following the instructions on page 17. Drink 1 cup (237 ml) a day, or as needed to relieve migraines.

VIOLET VINEGAR

Herbal vinegars are all the rage these days, and violet vinegar is especially flavorful. Use it as a base for salad dressing, drizzle a few drops over fruit, or splash it on your face as an after-cleansing aromatic toner.

3 ounces (84 g) violet flowers
1 quart (950 ml) high-quality white wine vinegar

Combine the flowers and the vinegar. Cover and let sit for several weeks in a warm place, shaking daily. Strain the herb using cheesecloth, then bottle. Store in a cool, dark location, where the vinegar will keep for up to 2 years.

CRYSTALLIZED VIOLETS

This is a great way to get more vitamin C, and it's fun, too!

10–20 violet flowers
Powdered egg-white mix
Fine castor sugar

1. Gently wash the violet flowers, then lay on paper towels to dry.

2. Prepare the egg white according to the package directions so that you have an amount equal to about 1 egg white, beaten.

3. Using a fine brush, coat each flower with the egg white. Dust with fine castor sugar. Snip the stems and set aside in a warm place (but not in direct sunlight) until dry, usually about 24 hours.

4. Store the violets in a sealed jar between layers of paper. Keep the jar in a cool, dark location.

VIOLET ICE CUBES

Whether floating in a punch bowl or served up with chilled lemonade or iced tea, violet ice cubes will make any party a little more festive!

1½ cups (360 ml) sugar
½ cup (120 ml) water
½ cup (120 ml) grape juice
1 handful violet petals
Juice of 3 lemons
1 handful whole violet flowers

1. In a saucepan, combine the sugar, water, grape juice, and violet petals. Bring to a boil, reduce the heat, and simmer for 10 minutes. Then remove from the heat and let cool.

2. Strain, then stir in the lemon juice. Blend well, then pour into ice cube trays. Set one whole violet flower in the middle of each filled cube compartment, then freeze.

VIOLET MOUSSE

This delicate dessert is perfect after a big meal. It refreshes the palate while providing a little extra shot of vitamins and minerals.

1 teaspoon (5 ml) gelatin or agar
¼ cup (60 ml) water
Powdered egg white mix
2 cups (470 ml) milk
2 tablespoons (30 ml) sugar
1 cup (235 ml) whipping cream
½ teaspoon (3 ml) violet tincture
Violet flowers

1. In a saucepan, combine the gelatin and water. Bring to a boil and cook for 2 minutes. Remove from the heat.

2. Prepare the egg white according to the package directions, making enough to equal about three egg whites, stiffly beaten.

2. When the gelatin mixture has cooled slightly, stir in the milk. Fold in the stiffly beaten egg whites and sugar, stirring until thick. Mix in the cream and violet tincture.

3. Pour into a mold and refrigerate until set. Garnish with violet flowers.

VIOLET HONEY

Violet honey was very popular in Victorian England as a spread and a flavoring for sauces. The violets give honey color, fragrance, flavor, and body. Violet honey can be used to sweeten tea and punches or as a sugar substitute in jellies, salad dressings, and frostings. It's delicious when blended with yogurt.

1 cup (235 ml) honey, room temperature
½ cup (120 ml) violet leaves and flowers, bruised

1. Put the honey in the top of a double boiler. Stir in the violets, cover, and cook for 10 minutes over low heat. (High heat damages honey, so don't let it boil.)

2. Remove from the heat, pour into sterilized jars, and seal. Let sit for 7 days at room temperature so the flavors will blend. Then rewarm over low heat, strain the herbs, and rebottle. Store in a cool, dark location, where the honey will keep for at least 2 years.

VIOLET SYRUP

Why limit yourself to maple syrup? Violet syrup is excellent on pancakes, waffles, and French toast. This recipe requires a lot of time (and a lot of flowers), but it's worth it.

1 quart (950 ml) boiling water
3 quarts (8.5 l) or more of violet flowers, packed
Honey

1. Pour the boiling water over 1 quart (950 ml) of compressed violet flowers. Let stand for 10 hours, then strain.

2. Reheat the liquid, bring to a simmer, and pour over a fresh batch of flowers. Let stand for another 10 hours, then strain.

3. Repeat again (and as often as you have flowers to supply — more is better).

4. When you run out of flowers, add honey to reach the proper syrup consistency. Store in a cool, dark location, where the syrup will keep for up to 1 year.

VIOLET SYRUP WITH SUGAR

If you don't have zillions of violet blossoms and unlimited time, this is the recipe you want.

1 cup (235 ml) water
3 cups (700 ml) sugar
1½ cups (360 ml) violet blossoms

In a nonreactive saucepan, boil the ingredients for 10 minutes, or until thickened. Strain through cheesecloth into a clean glass jar. Seal and store in the refrigerator. The syrup will keep for 2 to 4 weeks.

Variation. Not only is violet syrup good on pancakes, but it also makes a good drink. Pour 2 tablespoons (30 ml) of syrup over two or three ice cubes in a tall glass. Top with water and stir.

Violet Sherbet

To make a violet sherbet, stir violet syrup into newly fallen snow until the color and consistency are pleasing.

Violet Salad

With their peppery taste, violet leaves make a great addition to salads. Tear and toss together leaves of violets, butter lettuce, salad burnet, and endive. Dress with olive oil and marsala wine, garnish with violet blossoms, and serve.

CHILDREN'S COUGH SYRUP

This formula is good for sore throats as well as coughs, and children love the taste.

2 cups (470 ml) water
2 tablespoons (30 ml) fresh violet leaves and flowers, or 1 tablespoon (15 ml) dried
2–4 tablespoons (30–60 ml) honey

1. Bring the water to a boil. Stir in the violets, cover, and simmer for 20 to 30 minutes.

2. Stir in the honey, remove from the heat, and let cool. Take 1 to 2 tablespoons (15–30 ml) every morning and evening as needed. The syrup will keep for 2 to 4 weeks in the refrigerator.

VIOLET JAM

Because this wonderful jam doesn't require cooking, you'll get its great taste along with the maximum amount of nutrients.

1 cup (235 ml) violet flowers, tightly packed
1½ cups (360 ml) water, divided
Juice of 1 lemon
2½ cups (590 ml) sugar
1 2-ounce (56 g) package powdered pectin

1. Add the flowers, ¾ cup (180 ml) of the water, and the lemon juice to an electric blender. Blend into a smooth paste. Slowly add the sugar and blend until dissolved.

2. Stir the pectin into the remaining ¾ cup (180 ml) of water and bring to boil. Boil hard for 1 minute. Pour the cooked pectin into the blender. Blend for about 1 minute, then spoon into jars and seal tightly.

This jam keeps for about 3 weeks in the refrigerator. For longer storage, put it in the freezer.

VIOLET WATER

For a long time, violet water was a popular gift, especially for birthdays and saints' days. It's a tradition worth reviving. The aromatic water can be used as a fragrant refresher, like a splash.

- **2½ cups (590 ml) white wine vinegar**
- **1¼ cups (300 ml) distilled water**
- **4 cups (950 ml) violet flowers**

1. Combine the vinegar and water in a saucepan and bring to a boil. Pour over the flowers. Stir, cover, and let sit in a warm place for 3 weeks, stirring daily.

2. Strain through a nonmetal strainer, pressing the flowers to release more fragrance. Pour into bottles and store in a cool, dark location, where it will keep for up to 1 year.

VIOLET MOUTHWASH

This formula is recommended for tooth, gum, and mouth problems. The addition of sage is said to help whiten the teeth.

- **1 part chamomile flowers**
- **1 part comfrey leaves**
- **1 part lemon balm leaves**
- **1 part rosemary leaves**
- **1 part sage leaves**
- **1 part sorrel leaves**
- **1 part thyme leaves**
- **1 part violet flowers and leaves**

Make an infusion of the ingredients following the instructions on page 17. Let cool, then use as a mouthwash as needed. Store the left-overs in the refrigerator, where they will keep for 1 to 2 weeks.

VIOLET-SAGE CREAM

This cream helps relieve cold sores and soothes and protects chapped lips.

- **2 tablespoons (30 ml) finely chopped fresh sage**
- **2 tablespoons (30 ml) finely chopped sweet violet leaves**
- **8 tablespoons (120 ml) almond oil, divided**
- **4 tablespoons (60 ml) melted beeswax**

1. Mix the sage and sweet violet leaves with 4 tablespoons (60 ml) of the almond oil in a small jar. Seal and leave in a warm place for about 1 month, shaking each day.

2. Combine the remaining 4 tablespoons (60 ml) of almond oil with the melted beeswax in the top of a double boiler. Blend.

3. Strain the infused liquid and combine with the melted fats. Mix until cool and thick. Store in an airtight jar in a dark, cool place, where it will keep for 1 to 2 years. Apply twice a day, or as needed.

Violet Facial Steam and Wash

There's a good reason violet was once a popular ingredient in cosmetics: Its delicate floral scent is refreshing and not at all overpowering.

Make an infusion of leaves and flowers following the directions on page 17. Pour the infusion into a large bowl. Drape a towel over your head and lean over the bowl, allowing the steam to surround your face. If it's too hot, vent the towel to release some of the steam.

After steaming, strain the herb and use the remaining tea as a face wash.

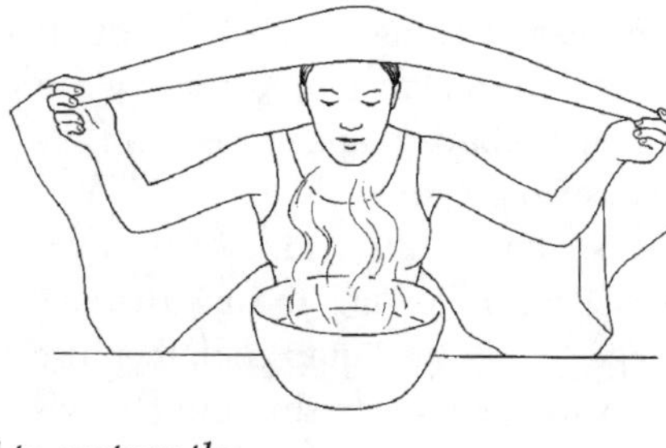

"Tent" a towel over both your head and the bowl to capture the aromatic steam.

VIOLET OINTMENT

This herbal remedy has been used for generations to heal sores — not only for people, but for animals as well.

1 ounce (28 g) lanolin
3 ounces (84 g) cocoa butter
Fresh violet leaves

1. Preheat the oven to 350°F (177°C).

2. Melt the lanolin and cocoa butter in a small ovenproof pot or jar. Add as many fresh violet leaves as the melted fats will cover. Bake in the oven for 1 hour.

3. Strain, then scoop the ointment into glass jars and seal tight. Store in a cool, dark location, where the ointment will keep for up to 5 years.

PLASTER FOR SORE NIPPLES

For breast-feeding mothers or women with dry skin, this poultice will give quick relief.

Boiling water
2 tablespoons (30 ml) crushed fresh violet leaves and flowers

Add enough boiling water to the crushed herbs to form a paste. Place between layers of gauze or cheesecloth and apply to the skin as needed.

Note: The plaster will retain its heat longer when mixed with bread or bran.

Baby Bottom Wash

Rather than washing your baby's bottom with plain water or dusting it with commercial products, try using this formula. It's gentle and helps relieve diaper rash. And it's very easy to prepare.

Make an infusion of violet leaves and flowers following the instructions on page 17. Let the liquid cool, then use it to wash and refresh your baby's bottom.